新手入廚系列

爛出美味

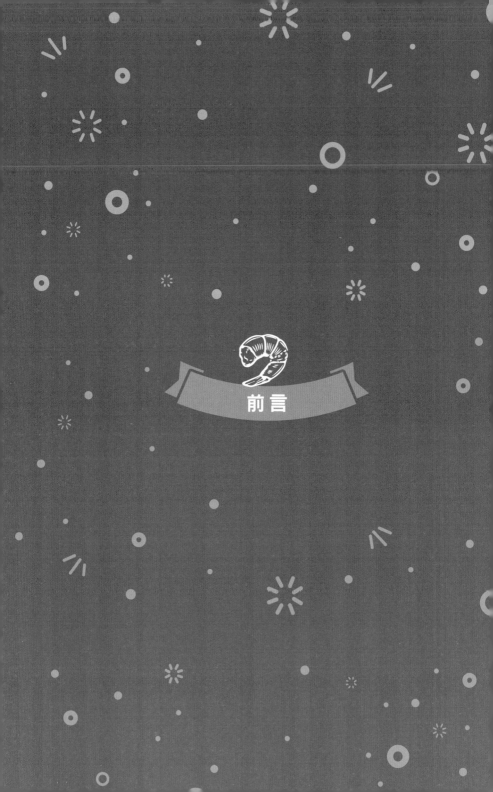

前言

都市人生活繁忙，未必有時間為家人預備「住家靚餸」。採用燜煮烹調方法，步驟非常簡單，便能把食材、湯汁融合為一，做出食材酥軟、原汁原味、醬汁豐厚的菜式，適宜長時間燜煮的食材。若未即時吃完，留至第二天，餸菜更為入味。

只要簡單地將食材和調味料放入電鍋烹煮，可以保存食物的原汁原味；用時間掣更可不用「看火」，完成後可以按保溫掣，亦可直接上桌。而鑄鐵鍋更兼具快速傳熱、保溫力強且外型美觀的特點，是入廚愛好者的得力助手。

本書以燜煮為主題，精選三十多款美味佳餚如南乳燜齋、素菜黃金袋、蜜燒子薑鱔、冬菇海參燜鴨掌、南乳蓮藕燜豬手和蘿蔔燜牛腩等輕鬆上桌的料理。從今天起，下班後都可以為自己預備少油、少鹽的菜式，善待自己。

目錄

節瓜
瓜身有密集絨毛,兩端圓潤,中間修腰,帶有凋謝瓜花。

Chinese marrow
Melon body with dense hair, rounded at both ends and narrower in the middle, with a withered flower.

芋頭
輕身,表面沒有水分,肉色潔白,微紫絲均勻分佈。

Taro
Light in weight with no water, flesh is white, light purple silk evenly distributed.

生薑
薑皮帶點泥土,有年輪紋理,肥美飽滿。

Ginger
Some soil attached, with age rings, plump and full.

栗子
雄栗子呈三角,清甜爽脆;雌栗子呈渾圓,粉糯腍軟。

Chestnut
Male chestnut in triangular shape, sweet and crispy; female chestnut rounded in shape, soft.

馬鈴薯（薯仔）
橢圓黃肉，無發芽，無損傷或黑心。

Potatoes
Oval and yellow, no germination,
no damage and black heart.

蓮藕
藕孔均勻，肥厚幼細，肉色粉紅。

Lotus root
Lotus holes are even, flesh is pink,
thick and smooth.

蘑菇
圓潤飽滿，菇柄細小，沒有損傷。

Mushrooms
Round and full, mushroom stalk is
small, no damage.

泰國苦瓜
結實青綠，形狀修長飽滿，色澤鮮明。

Thai bitter gourd
Solid and green, body slender and
full, bright color.

慈菇
菇蒂粗壯挺直，菇形呈長橢圓，
沒有損傷。

Arrowhead
Mushroom stalk is thick and
straight, mushroom shaped
elongated oval, no damage.

海參
肉厚肥大，堅實，沒有糜爛。
Sea cucumber
Flesh is thick and firm, no erosion.

龍脷柳
肉厚完整，色澤鮮明。
Sole fillet
Flesh is thick and intact, sharp color.

沙白蜆
鮮活，肉質肥美，沒有泥沙。
White sand clams
Fresh, plump meat, no sediment.

門鱔
肉質結實肥厚，皮光肉滑。
Eel
Flesh is thick, firm and smooth,
shinny skin.

油麵筋
色澤金黃，沒有油膩味道。
Oil gluten
Golden color, no unpleasant oily smell.

豬手
色澤潔白，皮薄，沒有瘀傷。

Trotter
White color, thin skin, no bruising.

油炸鴨掌
色澤金黃，肉厚肥大，沒有損破。

Fried duck's feet
Golden color, flesh is thick, no damage
nor broken.

這樣貯存好食材 Storage of ingredients

1. 鮮肉買回來不要清洗，立即用保鮮紙或保鮮盒包封，再放入冰格
   貯存。
2. 馬鈴薯、芋頭或南瓜毋須清洗，放在筲箕內，置陰涼地方貯放，
   可保存約 1~2 星期。
3. 乾貨如冬菇或粉絲放在密實盒，保持乾爽，可保存數月至一年。

1. Do not wash the meat bought and immediately wrap with plastic
   wrap or put into a plastic box, then place in the freezer.
2. No need to rinse root vegetables such as potatoes, taro or
   pumpkin, store in a strainer in a cool place, can store for 1 to
   2 weeks.
3. It is necessary to keep dried food such as dried black mushrooms
   or green bean noodles in a compact box, keep dry and it can be
   kept through the year.

市面發售的栗子有去殼或有殼的，公栗子扁平，母栗子渾圓肥大。

There are shelled and unshelled chestnuts in the market, male chestnuts are flat while female chestnuts are rounded and bigger.

剝栗子衣
Remove chestnut skin

1. 先放在滾水泡煮片刻。
2. 再趁熱用小刀立即去衣。

1. Cook chestnuts in boiling water for a while.
2. Remove the skin with a small knife while still hot.

註 Note
- 栗子凍了，不能去衣，如遇這情況，可再浸熱栗子處理，或是放微波爐加熱片刻。
- It is difficult to remove skin from chestnuts if they are cold. Dip chestnuts in hot water again or heat in the microwave for a while.

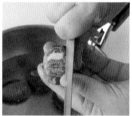

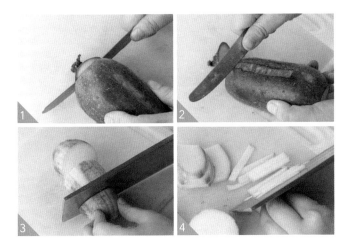

節瓜切粗絲
Shred Chinese marrow

1. 節瓜用小刀去掉首末兩端。
2. 刮掉外皮。
3. 切成 3 段。
4. 再切成粗絲。

1. Remove both ends with a small knife.
2. Scrape the skin.
3. Cut into 3 sections.
4. Then cut into thick shreds.

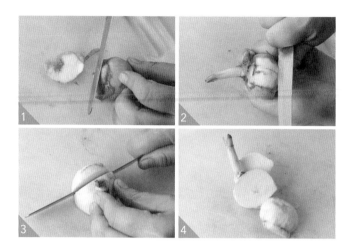

切慈菇
Cut arrowhead

1. 切去慈菇底部。
2. 用小刀刮掉外皮。
3. 切厚片,保留菇蒂完整形狀。
4. 一個慈菇約可切 3~4 片。

1. Cut the bottom of the arrowhead.
2. Scrape the skin with a small knife.
3. Cut into thick slices, reserve the shape of the arrowhead.
4. Cut each arrowhead into 3-4 slices.

爛餸怎樣才好味？ How to make tasty stewing dish?

原理 Principle

將食材放入砂鍋，加適量的湯水和調味料。蓋緊鍋蓋，煮滾後改用中小火進行較長時間的加熱，待食材酥軟入味，保留少量味汁便可。

Put ingredients into the casserole, add right amount of seasoning and soup. Tightly cover the lid and boil, switch to medium heat to cook for a long period until the ingredients are soft and tasty, retain less sauce.

優點 Advantages

物料軟爛，原汁原味，醬汁豐厚。若未即時吃完，留至第二天，更為入味。

Ingredients are soft, rich sauce retaining the original flavor. The dish will be more tasty if it is served the following day.

真假豆腐

True and False Beancurd

材料 | Ingredients

玉子豆腐 1 條
滑豆腐 1 磚
冬菇 3-4 朵
甘筍數片
蒜茸 1 茶匙
薑茸 1 茶匙

1 strip egg beancurd
1 brick soft beancurd
3 4 dried black mushrooms
a few slices of carrot
1 tsp minced garlic
1 tsp minced ginger

烟・蔬菜類

Stew・Vegetables

調味料 | Seasonings

糖 1 茶匙	1 tsp sugar
生粉 1 茶匙	1 tsp cornstarch
鹽 1/2 茶匙	1/2 tsp salt
老抽 1/2 茶匙	1/2 tsp dark soy sauce
清水 3 湯匙	3 tbsps water

做法 | Method

1. 板豆腐和玉子豆腐分別切件。熱鑊下 1-2 湯匙油，下豆腐煎至金黃，盛起瀝油。
2. 冬菇用生粉撈勻，沖洗乾淨，擠乾水分。燒熱鑊，用油炸冬菇片刻，瀝油。
3. 熱鍋下 1-2 茶匙油，放入蒜茸和薑茸爆香，放入冬菇、玉子豆腐和滑豆腐炒勻。
4. 下調味料煮至濃稠，再加入甘筍煮片刻，上碟。

1. Cut soft beancurd and egg beancurd into pieces. Heat 1-2 tbsps of oil in wok and pan-fry beancurd until golden brown, dish up and drain.
2. Scrub dried black mushrooms evenly with cornstarch, rinse, squeeze excess water. Deep-fry in wok for a while and drain.
3. Heat wok with 1-2 tsps of oil, sauté minced garlic and ginger until fragrant, add dried black mushrooms, egg beancurd and soft beancurd and stir well.
4. Add seasonings and cook until sauce thickens, add cooked carrots and cook for a while, serve.

入廚貼士 | Cooking Tips

- 煎豆腐時用易潔鍋容易做到較佳效果。
- It is easier to make this dish if pan-fry beancurd by using a non-stick wok.

南乳燜齋

Stewed Vegetables with Red Beancurd

材料 | Ingredients

黃芽白 600 克	600 g Tientsin cabbage
甜腐竹 75 克	75 g sweet beancurd sheets
腐竹 75 克	75 g beancurd sheets
粉絲 20 克	20 g green bean vermicelli
冬菇 16 朵（浸軟）	16 dried black mushrooms (soaked)
金針 10 克（浸軟）	10 g dried lily (soaked)
雲耳 10 克（浸軟）	10 g black fungus (soaked)
慈菇 4-5 個	4-5 arrowheads
南乳（小）1 磚	1 brick red beancurd (small)
薑 1 大塊（略拍）	1 large piece ginger (slightly pat)
清水 2 杯	2 cups water

4~6 人
Serves 4~6

10~15 分鐘
10~15 minutes

調味料 | Seasonings

片糖 1/2 片   1/2 piece slab sugar
鹽 1/4 茶匙   1/4 tsp salt

做法 | Method

1. 黃芽白洗淨，切段。慈菇刮皮，切片。

2. 冬菇、金針和雲耳用生粉撈勻，洗淨，擠乾。

3. 燒熱油鍋，下粉絲、慈菇、冬菇、甜竹、金針、雲耳和腐竹炸片刻，盛起瀝油。

4. 熱鍋下 1-2 茶匙油，放入薑片和南乳爆香，放入冬菇、金針、雲耳和黃芽白炒勻。

5. 注入清水煮至腍軟，放入粉絲、甜竹和腐竹焗片刻，下調味煮至濃稠，上碟。

1. Rinse Tientsin cabbage, cut into sections. Peel arrowheads and slice.

2. Scrub dried black mushrooms, dried lily and black fungus with cornstarch for a while, rinse, squeeze excess water.

3. Deep-fry green bean vermicelli, arrowheads, dried black mushrooms, sweet beancurd sheets, dried lily, black fungus and beancurd sheets for a while, dish up and drain.

4. Heat wok with 1-2 tsps of oil, sauté ginger slices and red beancurd until fragrant, add dried black mushrooms, dried lily, black fungus and Tientsin cabbage and stir well.

5. Add water and cook until soft, add green bean vermicelli, sweet beancurd sheets and beancurd sheets and cook for a while, add seasoning and cook until sauce thickens, serve.

入廚貼士 | Cooking Tips

- 各式材料必須經油炸後可保持完整。
- Deep-frying could keep all kinds of ingredients intact.

甜腐竹草菇爛豆腐

Stewed Beancurd with Sweet Beancurd Sticks and Straw Mushrooms

⬭⬭ 材料 | Ingredients

豆腐 1 磚（切件）　　甘筍數片
草菇 150 克　　　　上湯 1/2 杯
荷蘭豆 75 克（焯熟）　薑茸 1 茶匙
甜腐竹 40 克（炸透）　蒜茸 1 茶匙

1 brick beancurd (cut into pieces)
150 g straw mushrooms
75 g snow peas (blanched)
40 g sweet beancurd sheets (deep-fried thoroughly)
a few pieces of carrots
1/2 cup chicken stock
1 tsp minced ginger
1 tsp minced garlic

調味料 | Seasonings

糖 1 茶匙	1 tsp sugar
生粉 1 茶匙	1 tsp cornstarch
鹽 1/2 茶匙	1/2 tsp salt
清水 3 湯匙	3 tbsps water

做法 | Method

1. 鮮草菇削去底部，洗淨，開邊。放沸水中焯 1-2 分鐘，取出過冷，瀝乾。
2. 燒油一鍋至冒大白煙，放入豆腐以大火炸至金黃，盛起瀝油。
3. 熱鍋下 1-2 茶匙油，放入蒜茸和薑茸爆香，放入甜腐竹和油炸豆腐炒勻。加入上湯煮至腍軟。
4. 下調味料煮至濃稠，倒入草菇、荷蘭豆和甘筍片炒勻，上碟。

1. Cut the bottom of straw mushrooms, rinse and cut into halves. Blanch for 1-2 minutes, remove and rinse, drain.
2. Bring 1/2 wok of oil to the boil till white smoke appears, add beancurd and deep-fry till golden, dish up and drain.
3. Heat wok with 1-2 tsps of oil, sauté minced garlic and ginger until fragrant, add sweet beancurd sheets and fried beancurd and stir well. Add chicken stock and cook until soft.
4. Add seasoning and cook until sauce thickens, add straw mushrooms, snow peas and carrot slices and stir well, serve.

入廚貼士 | Cooking Tips

- 各式材料必須經油炸後才可保持完整。
- To keep all kinds of ingredients intact, deep-frying could help.

酸甜汁燜麵筋

Stewed Gluten in Sweet and Sour Sauce

材料 | Ingredients

菠蘿 2 片
油麵筋 1 包
三色椒各 1/2 個
青瓜 1/2 條（切片）
蒜茸 1 茶匙

2 slices pineapple
1 pack gluten
1/2 of each assorted color of bell peppers
1/2 cucumber (sliced)
1 tsp minced garlic

4~6 人
Serves 4~6

10~15 分鐘
10~15 minutes

◯◯◯ 酸甜汁 | Sweet and sour sauce

茄汁 1/3 杯	1/3 cup ketchup
白醋 3 湯匙	3 tbsps white vinegar
片糖 1 塊	1 slice slab sugar
喼汁 1 湯匙	1 tbsp Worcestershire sauce
老抽 1 茶匙	1 tsp dark soy sauce
生粉 1 茶匙	1 tsp cornstarch
鹽 1/2 茶匙	1/2 tsp salt
清水 1/3 杯	1/3 cup water

◯◯◯ 做法 | Method

1. 油麵筋放沸水焯 1 分鐘，撈出過冷。
2. 青瓜洗淨，切片。
3. 三色椒、菠蘿分別切角或切粒。燒熱鍋，用少許油炒片刻盛起。
4. 熱鍋下 1 茶匙油，放入酸甜汁和油麵筋煮至變稠，試味，熄火。再放入三色椒、菠蘿和青瓜片拌勻，盛起。

1. Blanch gluten in boiling water for 1 minute, remove and rinse.
2. Rinse cucumber and slice.
3. Cut assorted color of bell peppers and pineapple into wedges or dice respectively. Heat wok and stir-fry bell peppers with some oil and dish up.
4. Heat wok with 1 tsp of oil, add sweet and sour sauce and gluten and cook until sauce thickens, taste and switch off heat. Add peppers, pineapple and cucumber slices, mix well, dish up.

入廚貼士 | Cooking Tips

- 酸甜汁可因應個人調校味道。
- Sweet and sour sauce can be adjusted according to personal taste.

素菜黃金袋

Vegetarian Gold Bags

⊂⊃⊂⊃ 材料 | Ingredients

日本豆腐泡（油揚）4 件　　素翅 2 湯匙
竹笙 5 條　　　　　　　　蒜茸 1 茶匙
韭菜花 4 條　　　　　　　鹽 1/2 茶匙
鮮冬菇 2 朵　　　　　　　清雞湯 1/2 杯
蘆筍 2 條　　　　　　　　清水 1/2 杯
金菇適量

4 Japaness beancurd puffs
5 sticks bamboo fungus
4 stalk chives
2 fresh black mushrooms
2 asparagus
Some enoki mushrooms
2 tbsps vegetarian shark's fin
1 tsp minced garlic
1/2 tsp salt
1/2 cup chicken stock
1/2 cup water

調味料 | Seasonings

蠔油 1 湯匙	1 tbsp oyster sauce
生粉 2 茶匙	2 tsps cornstarch
糖 1/4 茶匙	1/4 tsp sugar
清水 3 湯匙	3 tbsps water

做法 | Method

1. 竹笙浸開後修剪頭尾，汆水，瀝乾，切粒。
2. 鮮冬菇、蘆筍、金菇分別洗淨，切粒；韭菜花汆水備用。
3. 燒熱 1 湯匙油，爆香蒜茸，放入鮮冬菇、蘆筍、金菇，竹笙和素翅炒片刻，下鹽調味。
4. 日本豆腐泡剪開口，釀入已炒好的材料；然後用韭菜花綁好。
5. 煮熱上湯，放入已釀好的豆腐泡煮 5 分鐘。加入芡汁煮至濃稠，便可上碟。

1. Trim heads and tails of bamboo fungus, blanch, drain and dice.
2. Rinse and dice fresh black mushrooms, asparagus and enoki mushrooms; blanch chives and set aside.
3. Heat 1 tbsp of oil, sauté minced garlic, add fresh black mushrooms, asparagus, enoki mushrooms, bamboo fungus and vegetarian shark's fin and stir-fry for a while, season with salt.
4. Cut opening slits on Japaness beancurd puffs, stuff stir-fried ingredients and tie with chives.
5. Bring chicken stock to the boil, add stuffed Japanese beancurd puffs and cook for 5 minutes. Add gravy and cook until thickens, dish up.

入廚貼士 | Cooking Tips
- 日本的油揚會比本地的豆腐泡適合做這道菜。
- Japaness beancurd puffs are more suitable than local beancurd puffs for making this dish.

粉絲蝦乾燜節瓜

Stewed Chinese Marrow with Green Bean Vermicelli and Dried Shrimps

材料 | Ingredients

節瓜 2 個（約 450 克）
蝦乾 20 克（用過面水浸軟）
粉絲 40 克（浸軟）
薑 1 片
清水 2 杯

2 Chinese marrows (about 450 g)
20 g dried shrimps (soak in water until soft)
40 g green bean vermicelli (soaked)
1 slice ginger
2 cups water

4~6 人
Serves 4~6

20~25 分鐘
20~25 minutes

⚮ 調味料 | Seasonings

糖 1 茶匙	1 tsp sugar
生粉 1 茶匙	1 tsp cornstarch
鹽 1/2 茶匙	1/2 tsp salt
清水 4 湯匙	4 tbsps water

⚮ 做法 | Method

1. 節瓜用小刀刮皮，洗淨，切絲。
2. 熱鑊下 1-2 湯匙油，放入薑片和蝦乾爆香，潷酒，倒入節瓜絲和清水煮 20 分鐘。
3. 加入粉絲和調味煮稠，上碟。

1. Peel Chinese marrows with a small knife, rinse and shred.
2. Heat wok with 1-2 tbsps of oil, sauté ginger slice and dried shrimps until fragrant, drizzle wine. Add Chinese marrow shreds and water and cook for 20 minutes.
3. Add green bean vermicelli and seasonings and cook until thickens, serve.

入廚貼士 | Cooking Tips

* 粉絲很吸水的，所以需要看情況下清水。
* Green bean vermicelliare very absorbent, so may need to adjust the amount of water.

雜菜煲 Mixed Vegetable Pot

◯◯◯ 材料 | Ingredients

蓮藕 200 克
草菇 40 克
蝦乾 20 克（用過面水浸軟）
甘筍數片
中國芹菜 1 棵（切段）
薑 1 片
上湯 1 杯

200 g lotus roots
40 g straw mushrooms
20 g dried shrimps (soak in water until soft)
a few pieces of carrot
1 Chinese celery (sectioned)
1 ginger slice
1 cup chicken stock

調味料 | Seasonings

鹽 1/2 茶匙
1/2 tsp salt

入廚貼士 | Cooking Tips

- 任何雜菜均可弄此菜式，最適合火鍋後翌日做這道菜。
- Any vegetables can be the used for this dish, it is most suitable to do this the day after serving hot pot.

做法 | Method

1. 草菇削去底部，洗淨，開邊後放沸水中焯煮 1 分鐘。
2. 蓮藕刮皮，洗淨，切薄片。
3. 熱鑊下 1 茶匙油，放薑片和蝦乾爆香，加入雜菜炒片刻，注入上湯煮沸。
4. 下調味料調勻，加入中國芹菜段拌勻便可。

1. Peel the bottom of straw mushrooms, rinse. Blanch in water for 1 minute.
2. Peel lotus roots, rinse and cut into thin slices.
3. Heat wok with 1 tsp of oil, sauté ginger slices and dried shrimps until fragrant, add mixed vegetables and stir-fry for a while, pour in chicken stock and bring to a boil.
4. Add seasonings and mix thoroughly, add Chinese celery and mix well, serve.

雙竹燜冬菇

Stewed Black Mushrooms with Double Beancurd Sticks

⚬⚬⚬ 材料 | Ingredients

小棠菜（上海白菜）300 克
鮮支竹 200 克（切段）
甜腐竹 75 克
冬菇 5-6 朵（浸軟）
上湯 1 杯
薑茸 1 茶匙

300 g Pak Choi (Shanghai cabbage)
200 g fresh beancurd sticks (sectioned)
75 g sweet beancurd sheets
5-6 dried black mushrooms (soaked)
1 cup chicken stock
1 tsp minced ginger

4~6 人
Serves 4~6

10~15 分鐘
10~15minutes

調味料 | Seasonings

糖 1 茶匙	1 tsp sugar
生粉 1 茶匙	1 tsp cornstarch
鹽 1/2 茶匙	1/2 tsp salt
清水 3 湯匙	3 tbsps water

做法 | Method

1. 甜腐竹用濕布抹乾。燒熱鍋，放大滾油以中火炸至金黃。
2. 冬菇用生粉撈勻，沖洗乾淨，擠乾水分。燒熱鍋，用油炸片刻，瀝油。
3. 小棠菜洗淨，以大火焯熟，瀝乾，放碟旁。
4. 熱鍋下 1-2 茶匙油，放入薑茸爆香，放入冬菇、甜腐竹和鮮支竹炒勻，加入上湯煮腍軟。下調味料煮至濃稠，上碟。

1. Wipe dry sweet beancurd sheets with a damp cloth. Heat wok and deep-fry over medium heat until golden.
2. Scrub dried black mushrooms evenly with cornstarch, rinse, squeeze excess water. Deep-fry for a while and drain.
3. Rinse Pak Choi. Heat wok and blanch over high heat, arrange onto a plate.
4. Heat wok with 1-2 tsps of oil, sauté minced ginger until fragrant, add dried black mushrooms, sweet beancurd sheets and beancurd sticks and stir well, add chicken stock and cook until soft. Add seasonings and cook until sauce thickens, serve.

入廚貼士 | Cooking Tips

- 冬菇先清洗再浸水，浸冬菇水保留作煮汁水用，可添加風味。
- Rinse dried black mushrooms first and then soak, water for soaking can be retained for cooking broth, hence adding flavor.

燜春菜

Stewed Haruna

⬤⬤⬤ 材料 | Ingredients

春菜 600 克
豬腩肉 1 條（約 225 克）
蒜子 3 粒
乾蔥頭 2 粒
薑 2 片
豆醬 2 湯匙

600 g haruna
1 strip pork belly (approximately 225 g)
3 cloves garlic
2 shallots
2 slices ginger
2 tbsps fermented bean sauce

⑪ 醃料 | Marinade

鹽 1/2 湯匙
1/2 tbsp salt

入廚貼士 | Cooking Tips

- 菜梗和菜葉分開烹煮，因其烹調時間不同，還可確保春菜嫩綠色澤。
- To keep haruna green in color, cook stems and leaves in different duration.

⑪ 做法 | Method

1. 豬腩肉用鹽撈勻，醃 1/2 小時，切片。
2. 春菜切成 1 吋長段，而梗和葉要分開擺放。
3. 燒熱鍋，將乾葱頭、蒜子、薑先用油爆香，加入豬腩肉煎至金黃色。
4. 放進春菜的梗部先煮 2 分鐘，再加入春菜葉、豆醬續爛煮 1/2 小時，上桌。

1. Marinate pork belly with salt for 1/2 hour, then slice.
2. Cut haruna into1-inch long sections, separate stems and leaves.
3. Heat wok and sauté shallots, garlic and ginger with oil until fragrant, add pork belly and pan-fry until golden brown.
4. Add stems of haruna and cook for 2 minutes, then add leaves and fermented bean sauce and cook for 1/2 hour. Serve.

肉碎茄子

Stewed Eggplant with Minced Pork

材料 | Ingredients

茄子 450 克
免治豬肉 150 克
蒜茸 1 茶匙
薑茸 1 茶匙
辣椒碎 1 茶匙

450 g eggplant
150 g minced pork
1 tsp minced garlic
1 tsp minced ginger
1 tsp minced red chili

4~6 人
Serves 4~6

30 分鐘
30 minutes

醃料 | Marinade

糖 1 茶匙
生粉 1 茶匙
生油 1 茶匙
鹽 1/2 茶匙
紹興酒 1/2 茶匙

1 tsp sugar
1 tsp cornstarch
1 tsp oil
1/2 tsp salt
1/2 tsp Shaoxing wine

芡汁 | Gravy

糖 2 茶匙
生粉 1 茶匙
老抽 1/2 茶匙
鹽 1/4 茶匙
清水 3 湯匙

2 tsps sugar
1 tsp cornstarch
1/2 tsp dark soy sauce
1/4 tsp salt
3 tbsps water

做法 | Method

1. 茄子去蒂，切條，浸鹽水待 5 分鐘，瀝乾。燒熱鍋，放大滾油中炸透，盛起瀝油。

2. 免治豬肉加醃料拌勻，待 5 分鐘。

3. 熱鍋下油，放入蒜茸、薑茸和辣椒碎爆香，加入免治豬肉炒至微黃。

4. 倒入茄子炒透，下芡汁煮稠便可。

1. Remove stalks from eggplant, cut into strips, soak in salt water for 5 minutes, drain. Heat wok and deep-fry in oil, dish up and drain.

2. Marinate minced pork for 5 minutes.

3. Heat oil in wok, add minced garlic, ginger and red chili and sauté until fragrant. Add minced pork and stir-fry until slighty yellowish.

4. Add eggplant and stir-fry well, add sauce and cook until thickens. Serve.

入廚貼士 | Cooking Tips

- 茄子用鹽水稍浸可去澀味，而必須用大滾油炸透方可保持色澤鮮艷，否則要把茄子皮削去。
- To remove the bitter taste of eggplant, soak in salt water for a while. In order to maintain the color of eggplant, it must be deep-fried in oil, or the skin of eggplant peeled away.

鯪魚肉釀豆泡煮蘿蔔

Stewed Mud Fish and Stuffed Beancurd Puffs with Turnip

🍳 材料 | Ingredients

白蘿蔔 300 克	300 g turnip
鯪魚肉茸 150 克	150 g minced mud fish
豆腐泡 10-12 個	10-12 beancurd puffs
中國蒜 1 條（切段）	1 Chinese garlic (sectioned)
芹菜 1 條（切段）	1 Chinese celery (sectioned)
清水 1/2 杯	1/2 cup water

🍳 醃料 | Marinade

糖 1 茶匙	1 tsp sugar
生粉 1 茶匙	1 tsp cornstarch
鹽 1/4 茶匙	1/4 tsp salt
麻油 1/4 茶匙	1/4 tsp sesame oil
胡椒粉適量	Pinch of pepper

4~6 人
Serves 4~6

30 分鐘
30 minutes

調味料 | Seasonings

生粉 1 茶匙
鹽 1/ 2 茶匙
糖 1/2 茶匙
清水 1 湯匙
1 tsp cornstarch
1/2 tsp salt
1/2 tsp sugar
1 tbsp water

入廚貼士 | Cooking Tips

- 處理鯪魚茸時，器皿用具不要沾有薑或蒜，否則魚肉會變霉和不結實，沒有彈性。
- When dealing with minced mud fish, the utensils should not stained with ginger or garlic, otherwise the fish will not be sticky and firm.

做法 | Method

1. 鯪魚肉加入醃料，以順時針拌至有膠質和結實，置冰箱冷藏 10 分鐘。

2. 白蘿蔔去皮，洗淨，切角。

3. 豆腐泡用水沖洗，切半。釀入鯪魚肉。

4. 熱鍋下油，放豆腐泡煎熟和至微焦，盛起。

5. 熱鍋下油，放入芹菜梗和中國蒜的白色部分爆香。放入白蘿蔔和清水煮至開始變透明。

6. 放入豆腐泡續煮至熟透和變軟，下調味料煮至濃稠。最後加入綠蒜葉拌勻。

1. Marinate minced mud fish, stir in clockwise direction until sticky and firm. Put into a freezer for 10 minutes.

2. Peel turnip, rinse and cut into wedges.

3. Rinse beancurd puffs, cut in halves. Stuff minced mud fish into beancurd puffs.

4. Heat oil in wok, pan-fry stuffed beancurd puffs till done and slightly brown, dish up.

5. Heat oil in wok, sauté Chinese celery stalks and white part of Chinese garlic until fragrant. Add turnip and water and cook until turnip starts to turn into transparent.

6. Add stuffed beancurd puffs and cook until done and soft, add seasonings and cook until sauce thickens. Add green leaves of Chinese garlic and mix well.

魚腐燜黃芽白

Stewed Deep-fried Fish Puffs with White Cabbage

材料 | Ingredients

津白 / 紹菜 600 克
魚腐 6-8 個
薑 2-3 片
蒜頭 1 粒（略拍）
清水 1/2 杯

600 g Tianjin white cabbage
6-8 deep-fried fish puffs
2-3 slices ginger
1 clove garlic (slightly pat)
1/2 cup water

◯◯ 調味料 | Seasonings

糖 1 茶匙	1 tsp sugar
蠔油 1 茶匙	1 tsp oyster sauce
生粉 1 茶匙	1 tsp cornstarch
鹽 1/2 茶匙	1/2 tsp salt
老抽 1/2 茶匙	1/2 tsp dark soy sauce
清水 3 湯匙	3 tbsps water

◯◯ 做法 | Method

1. 津白去掉老葉，切成長段。浸在水中 10 分鐘，再清洗 2 次，撈出後瀝水。
2. 魚腐放沸水中焯 1-2 分鐘，去掉肥油，切成兩半。
3. 熱鍋下 1-2 湯匙油，放入薑片爆香，倒入津白和清水煮 10 分鐘。
4. 加入魚腐煮 3-4 分鐘，下調味料煮至濃稠和汁略為收乾，上碟。

1. Remove old leaves from Tianjin white cabbage, cut into long sections. Soak in water for 10 minutes, then rinse for 2 times, drain.
2. Blanch deep-fried fish puffs in boiling water for 1-2 minutes, remove oil, cut in halves.
3. Heat wok with 1-2 tbsps of oil, add ginger slices and sauté until fragrant, add Tianjin white cabbage and water and cook for 10 minutes.
4. Add deep-fried fish puffs and cook for 3-4 minutes, add seasonings and cook until thickens, serve.

> 入廚貼士 | Cooking Tips
> - 沒有津白可改用娃娃菜或潮州白菜，這些菜葉含水分較少，可多添加點清水煮脸。
> - Tianjin white cabbage could be replaced by white cabbage or Chaozhou white cabbage, however, more water is needed as the leaves are less moisture.

魚湯煮蜆

Stewed Clars in Fish Soup

鮮雜魚 600 克
海蜆 600 克
草菇 3-4 粒（汆水開邊）
中國芹菜 1 棵（切段）
甘筍數片
清水 1 杯

600 g fresh mixed fishes
600 g sea clams
3-4 straw mushrooms (blanched and halved)
1 Chinese celery (sectioned)
some pieces of carrot
1 cup water

4~6 人
Serves 4~6

20~25 分鐘
20~25 minutes

調味料 | Seasonings

鹽 1/2 茶匙　　　　　1/2 tsp salt
胡椒粉少許　　　　　Pinch of pepper

做法 | Method

1. 鮮雜魚洗淨，徹底去除內臟，用少許鹽醃片刻。
2. 熱鍋下 2-3 湯匙油，放入雜魚煎至兩面微黃。下清水以大火煮至奶白色，隔渣留湯。
3. 海蜆洗淨，放滾水中煮至剛開口，撈出。
4. 把魚湯、草菇和甘筍片煮沸，放入海蜆和中國芹菜段煮沸便可。

1. Rinse and gut fresh mixed fishes, marinate with a pinch of salt.
2. Heat wok with 2-3 tbsps of oil, pan-fry fresh mixed fishes until both sides slighty yellowish. Add water and cook over high heat until the soup turns milky white, retain soup.
3. Rinse sea clams, cook in boiling water until shells just open, dish up.
4. Bring fish soup, straw mushrooms and carrot slices to a boil, add sea clams and Chinese celery sections and bring to a boil again. Serve.

入廚貼士 | Cooking Tips

- 煎魚時先把油鍋用慢火燒熱，下油，待冒煙時，放入魚。不要移動，輕搖鍋，待魚能離鍋，方可翻轉。如見黏底，可多下點油。
- When pan-frying fish, heat the wok with oil first and add fish when smoke forms. Do not move fish, shake the wok slightly until fish is away from the wok, then turns to another side. If the fish stick to the wok, some more oil could be added.

紅燒魚塊

Braised Fish Fillets

⟨⟨⟨ 材料 | Ingredients

龍脷魚 450 克　　　　　　薑 2-3 片
燒腩 300 克（切塊）　　　蒜頭 2-3 粒（拍扁）
冬菇 5-6 朵（浸軟）　　　葱 1 條
冬筍 40 克（汆水，切角）　清水 1 杯

450 g sole fish
300 g roasted pork (cut into pieces)
5-6 dried black mushrooms (soaked)
40 g bamboo shoots (blanched and cut into wedges)
2-3 slices ginger
2-3 cloves garlic (pat)
1 stalk spring onion
1 cup water

醃料 | Marinade

糖 1 茶匙	1 tsp sugar
生粉 1 茶匙	1 tsp cornstarch
鹽 1/4 茶匙	1/4 tsp salt
麻油 1/4 茶匙	1/4 tsp sesame oil
胡椒粉適量	Pinch of pepper

調味料 | Seasonings

糖 1 茶匙	1 tsp sugar
老抽 1 茶匙	1 tsp dark soy sauce
生粉 1 茶匙	1 tsp cornstarch
鹽 1/2 茶匙	1/2 tsp salt
清水 2 湯匙	2 tbsps water

做法 | Method

1. 魚腩洗淨，切塊，抹乾水分。加入醃料拌勻，撲少許生粉。燒熱鍋，用中慢火煎香。
2. 冬菇用少許生粉撈勻，洗淨，擠乾水分。
3. 熱鍋下 2 湯匙油，放入薑片、葱白、蒜頭和冬菇爆香，放入燒腩和冬筍，注入清水燜 15 分鐘。
4. 放入魚腩和餘下葱段燜 5 分鐘，下調味煮至汁濃稠。

1. Rinse sole fish, cut into pieces, drain. Add marinade and mix well, pat with some cornstarch. Heat wok and pan-fry over low to medium heat until fragrant.
2. Rinse dried black mushrooms with some cornstarch, squeeze excess water.
3. Heat wok with 2 tbsps of oil, add ginger slices, white part of spring onion, garlic and dried black mushrooms and sauté until fragrant, add roasted pork, bamboo shoots and water and stew for 15 minutes.
4. Add fish fillet and remaining spring onion and stew for 5 minutes, add seasonings and cook until sauce thickens.

入廚貼士 | Cooking Tips

- 炸魚腩前預先拌勻醃料，炸前才撲生粉；還要用手輕輕壓實，才不會在油炸時掉粉，弄污炸油，令炸魚不金黃和黏上污物。
- Before deep-frying fish fillet, mix well with marinade and then pat with cornstarch gently with hands, then cornstarch will not fall out into deep-frying oil easily and make fish fillet not golden in color and sticked with dirts.

鹹菜燜門鱔

Stewed White Eel with Pickled Cabbage

材料 | Ingredients

門鱔 450 克	450 g white eel
鹹酸菜 150 克（切塊）	150 g pickled cabbage (cut into pieces)
中國芹菜 1 條	1 Chinese celery
中國蒜 1 條	1 Chinese garlic
蒜頭 4-5 粒（拍扁）	4-5 cloves garlic (pat flat)
薑 2-3 片	2 or 3 slices ginger
清水 1/2 杯	1/2 cup water

醃料 | Marinade

紹興酒 1 湯匙	1 tbsp Shaoxing wine
糖 1 茶匙	1 tsp sugar
生粉 1 茶匙	1 tsp cornstarch
鹽 1/4 茶匙	1/4 tsp salt
麻油 1/ 4 茶匙	1/4 tsp sesame oil
胡椒粉適量	Pinch of pepper

4~6 人
Serves 4~6

20~25 分鐘
20~25 minutes

⟨⟨⟩⟩ 調味料 | Seasonings

糖 1 茶匙	1 tsp sugar
生粉 1 茶匙	1 tsp cornstarch
鹽 1/2 茶匙	1/2 tsp salt
清水 2 湯匙	2 tbsps water

入廚貼士 | Cooking Tips

- 中國芹菜和蒜可先用油炒至半熟，最後加入菜餚，保持青綠，味道好。
- China celery and garlic could be stir-fried until half-cooked and finally add to the dish to keep them green and tasty.

⟨⟨⟩⟩ 做法 | Method

1. 鹹酸菜用清水浸 30 分鐘，取出，用鹽擦洗，再過清水，然後切塊。燒熱鍋，用白鍋烘乾，加 1 湯匙糖撈勻，備用。

2. 門鱔洗淨，用少許鹽擦洗去潺，沖洗。再放微沸水中浸 1 分鐘，取出，用小刀刮去潺。

3. 抹乾水分，切塊。加醃料拌勻，撲少許生粉。燒熱鍋，用中慢火煎香。

4. 熱鍋下 2 湯匙油，放入薑片、中國蒜子頭和蒜頭爆香，加入白鱔和鹹酸菜後灒酒，下調味料，蓋鍋蓋焗煮 15-20 分鐘至熟。

5. 最後，加入芹菜和中國蒜炒勻，上碟。

1. Soak pickled cabbage in water for 30 minutes, remove, scrub with salt, rinse with water again, then cut into pieces. Heat a wok and sauté without adding oil, add 1 tbsp of sugar and mix well, set aside.

2. Rinse white eel and scrub with some salt to remove sticky dirts, rinse. Put white eel into slightly boiling water for 1 minute, remove and scrape with a small knife to remove sticky dirts.

3. Pat dry and cut into pieces, add marinade and mix well, pat with some cornstarch. Heat a wok and pan-fry over low to medium heat until fragrant.

4. Heat wok with 2 tbsps of oil, add ginger slices, head of garlic and sauté until fragrant, add white eel and pickled cabbage, drizzle wine, add seasonings. Cook with lid covered for 15-20 minutes.

5. Finally, add Chinese celery and garlic and stir well. Serve.

蒜子燜鱔

Stewed Eel with Garlic

⚬⚬⚬ 材料 | Ingredients

白鱔 450 克
燒腩 300 克（切塊）
冬筍 40 克（汆水、切角）
獨子蒜 10-12 粒
冬菇 5-6 朵（浸軟）

中國芹菜 1 條
中國蒜 1 條
薑 2-3 片
清水 1 杯

450 g white eel

300 g roasted pork (cut into pieces)

40 g bamboo shoots (blanched and cut into wedges)

10-12 garlic

5-6 dried black mushrooms (soaked)

1 Chinese celery

1 Chinese garlic

2-3 slices ginger

1 cup water

醃料 | Marinade

紹興酒 1 湯匙	1 tbsp Shaoxing wine
糖 1 茶匙	1 tsp sugar
生粉 1 茶匙	1 tsp cornstarch
鹽 1/4 茶匙	1/4 tsp salt
麻油 1/4 茶匙	1/4 tsp sesame oil
胡椒粉適量	Pinch of pepper

調味料 | Seasonings

糖 1 茶匙	1 tsp sugar
生粉 1 茶匙	1 tsp cornstarch
鹽 1/2 茶匙	1/2 tsp salt
清水 2 湯匙	2 tbsps water

做法 | Method

1. 白鱔洗淨，用少許鹽擦洗去潺，沖洗。再放微沸水中浸 1 分鐘，取出，用小刀刮潺。抹乾水份，切塊，加醃料勻，撲少許生粉。
2. 燒熱鍋，用中慢火煎香。
3. 蒜子去衣，燒熱鍋，用中火炸至金黃，取出備用。
4. 熱鍋下 2 湯匙油，放入薑片、蒜子頭和冬菇爆香；放入燒腩、冬筍和白鱔，注入清水燜 15-20 分鐘。
5. 下調味和已炸蒜子煮至汁濃稠，加入中國芹菜和蒜子拌勻，上碟。

1. Rinse white eel with some salt and scrub to remove sticky dirts, rinse. Put white eel into slightly boiling water for 1 minute, remove and scrape with a small knife to remove sticky dirts. Pat dry and cut into pieces, add marinade and mix well, pat with some cornstarch.
2. Heat wok and pan-fry over medium heat until fragrant.
3. Skin garlic, heat wok and deep-fry over medium heat until golden, dish up and set aside.
4. Heat wok with 2 tbsps of oil, add ginger slices, head of garlic and dried black mushrooms and sauté until fragrant. Add roasted pork, bamboo shoots and white eel, add water and stew for 15-20 minutes.
5. Add seasoning and deep-fry garlic and cook until sauce thickens, add Chinese celery and garlic and mix well, serve.

入廚貼士 | Cooking Tips

- 用已炸的乾葱頭和蒜頭的油煮菜，味道更香。
- The dish will be more tasty if cooked with the oil deep-frying shallots and garlic.

蜜燒子薑鱔

Stewed White Eel with Ginger in Honey

材料 | Ingredients

白鱔 450 克
酸子薑 10 克
紅椒 1 隻（切粒）
青瓜 1 條（約 150 克，切薄片和裝飾）
450 g white eel
10 g sour young ginger
1 red chilli (diced)
1 cucumber (about 150 g, cut into thin slices for garnish)

4~6 人
Serves 4~6

20 分鐘
20 minutes

醃料 | Marinade

紹興酒 1 湯匙	1 tbsp Shaoxing wine
糖 1 茶匙	1 tsp sugar
生粉 1 茶匙	1 tsp cornstarch
鹽 1/4 茶匙	1/4 tsp salt
麻油 1/4 茶匙	1/4 tsp sesame oil
胡椒粉適量	Pinch of pepper

蜜汁 | Honey sauce

檸檬汁 1 湯匙	1 tbsp lemom juice
糖 2 茶匙	2 tsps sugar
蜜糖 1 茶匙	1 tsp honey
清水 3 湯匙	3 tbsps water

做法 | Method

1. 白鱔洗淨，用少許鹽擦洗去潺，沖洗。
2. 放微沸水中浸 1 分鐘，取出用小刀刮潺，抹乾水份，切塊。加醃料勻，撲少許生粉，輕輕壓實。
3. 燒油一鍋至八成滾，放入鱔魚炸至金黃，撈出瀝油。
4. 熱鍋下油，倒進蜜汁煮沸，下白鱔魚煮至收汁，加入酸子薑和辣椒粒炒勻上碟。

1. Rinse white eel and scrub with some salt to remove sticky dirts, rinse.
2. Put white eel into slightly boiling water for 1 minute, drain and scrape with a small knife to remove sticky dirts. Pat dry and cut into pieces, add marinade and mix well, pat with some cornstarch. Press slightly.
3. Heat oil to 80% boil, add white eel and deep-fry till golden, drain.
4. Heat oil in wok, pour in honey sauce and bring to a boil, add white eel and cook until sauce dries up, add sour young ginger and chopped red chilli and stir well. Serve.

入廚貼士 | Cooking Tips
- 蜜糖不宜煮太久，否則會變酸。
- Honey should not be cooked too long or it will become sour.

栗子燜雞翼

Stewed Chicken Wings with Chestnuts

材料 | Ingredients

雞翼 450 克	450 g chicken wings
栗子 300 克	300 g chestnuts
薑 2-3 片	2-3 slices ginger
蒜頭 1-2 粒	1-2 cloves garlic
清水 1-2 杯	1-2 cups water

醃料 | Marinade

油 1/2 湯匙	1/2 tbsp oil
薑汁酒 2 茶匙	2 tsps ginger wine
鹽 1 茶匙	1 tsp salt
糖 1 茶匙	1 tsp sugar
生粉 1 茶匙	1 tsp cornstarch
老抽 1/2 茶匙	1/2 tsp dark soy sauce

4~6 人
Serves 4~6

20~25 分鐘
20~25 minutes

調味料 | Seasonings

蠔油 2 茶匙	鹽 1/4 茶匙	2 tsps oyster sauce
糖 1 茶匙	清水 4 湯匙	1 tsp sugar
生粉 1 茶匙		1 tsp cornstarch
		1/4 tsp salt
		4 tbsps water

做法 | Method

1. 雞翼解凍，用少許鹽擦洗，沖淨，抹乾水分。加入醃料醃 1 小時備用。
2. 栗子放沸水中焯 1 分鐘，取出趁熱去衣，洗淨。
3. 熱鍋下 1-2 湯匙油，放雞翼煎至微黃，盛起。
4. 熱鍋下 1-2 湯匙油，放入薑片和蒜頭爆香；倒入栗子炒數下，加入雞翼和清水煮 15-20 分鐘。
5. 下調味料煮至濃稠收汁，上碟。

1. Thaw chicken wings, scrub with some salt and rinse, pat dry. Add marinade and marinate for 1 hour. Set aside
2. Blanch chestnuts in boiling water for 1 minute, drain and remove skin while hot, rinse.
3. Heat wok with 1-2 tbsps of oil, pan-fry chicken wings till yellowish, dish up.
4. Heat wok with 1-2 tbsps of oil, add ginger slices and garlic and sauté until fragrant. Add chestnuts and stir-fry for a while, add chicken wings and water and cook for 15-20 minutes.
5. Add seasoning and cook until sauce thickens, serve.

入廚貼士 | Cooking Tips

- 雞翼用鹽擦洗，可把雪味去除。至於醃味時多點生粉會使雞肉比較嫩滑，但容易黏底，煎雞翼時要加倍注意。
- The smell of frozen could be removed if the chicken wings are scrubbed and rinsed with salt. When marinating, add more cornstarch will make chicken more tender. However, it will be easy to stick to the wok, we should pay more attention when pan-frying chicken wings.

豆醬燜雞翼
Stewed Chicken Wings with Chaozhou Bean Paste

材料 | Ingredients

雞翼 450 克	450 g chicken wings
薑茸 1 茶匙	1 tsp minced ginger
蒜茸 1 茶匙	1 tsp minced garlic
辣椒茸 1 茶匙	1 tsp minced red chilli
清水 1/2 杯	1/2 cup water

醃料 | Marinade

油 1/2 湯匙	1/2 tbsp oil
薑汁酒 2 茶匙	2 tsps ginger wine
鹽 1 茶匙	1 tsp salt
糖 1 茶匙	1 tsp sugar
老抽 1/2 茶匙	1/2 tsp dark soy sauce
生粉 1 茶匙	1 tsp cornstarch

調味料 | Seasonings

潮州豆醬 1 湯匙	1 tbsp Chaozhou bean paste
糖 2 茶匙	2 tsps sugar
生粉 1 茶匙	1 tsp cornstarch
鹽 1/4 茶匙	1/4 tsp salt
清水 5 湯匙	5 tbsps water

做法 | Method

1. 潮州豆醬用隔篩磨幼。
2. 雞翼解凍，用少許鹽擦洗，沖淨，抹乾水分。加入醃料醃 1 小時備用。
3. 熱鍋下 1-2 湯匙油，放雞翼煎至微黃，盛起。
4. 原鍋加熱，下調味料煮沸，放入雞翼煮 5 分鐘，即成。

1. Drain Chaozhou bean paste through a fine stainer to make it smooth.
2. Thaw chicken wings, scrub with some salt and rinse, pat dry. Add marinade and marinate for 1 hour. Set aside.
3. Heat wok with 1-2 tbsps of oil, pan-fry chicken wings till yellowish, dish up.
4. Heat wok, add seasonings and bring to a boil, add chicken wings and cook for 5 minutes and serve.

入廚貼士 | Cooking Tips

- 有骨的雞翼不易入味，所以加入醃料拌勻，必須待久一點，要不是就只有雞皮有味，而肉就沒味了。
- Chicken wings with bones are not easy to absorb marinade, so the time for marinating must be longer, otherwise only chicken skin is flavored but the meat has no flavor.

雞粒豆腐煲

Stewed Chicken Dice and Beancurd Pot

材料 | Ingredients

雞髀 1 隻
豆腐 1 盒
薑茸 1 茶匙
蒜茸 1 茶匙
辣椒茸 1 茶匙
葱 1 條（切粒）
芫荽 1 條（切粒）

1 chicken thigh
1 box beancurd
1 tsp minced ginger
1 tsp minced garlic
1 tsp red chilli
1 spring onion (diced)
1 coriander (diced)

4~6 人
Serves 4~6

10~15 分鐘
10~15 minutes

<table>
<tr><td colspan="2">醃料 | Marinade</td><td colspan="2">調味料 | Seasonings</td></tr>
</table>

醃料 | Marinade		調味料 | Seasonings	
糖 1 茶匙	1 tsp sugar	糖 1 茶匙	1 tsp sugar
薑汁酒 1 茶匙	1 tsp ginger wine	生粉 1 茶匙	1 tsp cornstarch
生粉 1 茶匙	1 tsp cornstarch	鹽 1/4 茶匙	1/4 tsp salt
油 1 茶匙	1 tsp oil	清水 3 湯匙	3 tbsps water
鹽 1/2 茶匙	1/2 tsp salt		

做法 | Method

1. 雞髀洗淨、去骨、抹乾、切粒,加入醃料拌勻,待 5-10 分鐘。
2. 豆腐切粒,放入沸水中焯 1-2 分鐘,連水盛起。
3. 熱鍋下 1-2 湯匙油,加入薑茸、蒜茸和辣椒茸爆香,倒入雞粒炒熟。
4. 豆腐盛起,瀝水,倒入雞粒內炒透。下調味料煮至濃稠,撒葱粒和芫荽粒,上碟。

1. Rinse chicken thigh, remove bones, pat dry and dice, add marinade and leave for 5-10 minutes.
2. Dice beancurd, blanch in boiling water for 1-2 minutes, dish up with water.
3. Heat wok with 1-2 tbsps of oil, add minced ginger, garlic and red chilli and sauté until fragrant, add chicken dice and stir-fry until done.
4. Dish up beancurd and drain, add chicken dice and stir-fry thoroughly. Add seasoning and cook until sauce thickens, sprinkle chopped spring onion and coriander. Serve.

入廚貼士 | Cooking Tips

- 豆腐焯太久會出現蜂巢狀,短時間焯煮比較合適,而原豆腐水的餘溫,繼續令豆腐保溫。
- Beancurd could not be cooked for too long or there will be honeycomb-like, cook for short while will be most desirable. Also, the warmth of beancurd water continues to keep beancurd warm.

涼瓜燜火鴨

Braised Duck with Bitter Gourd

材料 | Ingredients

涼瓜（苦瓜）1 個
燒鴨 300 克（切件）
薑茸 1 茶匙
蒜茸 1 茶匙
辣椒茸 1 茶匙
豆豉 1 茶匙
清水 1/2 杯

1 bitter gourd
300g roasted duck (cut into pieces)
1 tsp minced ginger
1 tsp minced garlic
1 tsp minced red chili
1 tsp fermented black beans
1/2 cup water

調味料 | Seasonings

糖 2 茶匙	2 tsps sugar
生粉 1 茶匙	1 tsp cornstarch
老抽 1/2 茶匙	1/2 tsp dark soy sauce
鹽 1/4 茶匙	1/4 tsp salt
清水 3 湯匙	3 tbsps water

做法 | Method

1. 涼瓜去籽，切塊，加少許鹽撈勻待片刻，沖水，瀝乾。
2. 熱鍋下 1-2 湯匙油，放入薑茸、蒜茸、辣椒茸和豆豉爆香。
3. 倒入火鴨和涼瓜塊炒透，潷酒，注入清水煮 10 分鐘。
4. 下調味料煮至濃稠收汁，上碟。

1. Seed bitter gourd, cut into pieces, add some salt and mix, leave for a while, rinse and drain.
2. Heat wok with 1-2 tbsps of oil, sauté minced ginger, garlic, red chili and fermented black beans until fragrant.
3. Add roasted duck and bitter gourd and stir-fry well. Drizzle wine, add water and cook for 10 minutes.
4. Add seasoning and cook until sauce thickens, dish up.

入廚貼士 | Cooking Tips

- 涼瓜生炒味道較佳和質感爽脆，但要用點鹽略醃，令其水分排出和降低苦味。
- Bitter gourd is tasty and crispy if it is stir-fried. Mix with some salt to discharge water and reduce bitterness.

荔芋香鴨煲

Taro and Duck Pot

材料 | Ingredients

鴨 1/2 隻	1/2 duck
荔芋 1/2 個	1/2 taro

調味料 | Seasonings

鹽 1 茶匙	1 tsp salt
糖 1 茶匙	1 tsp sugar
米酒 1 茶匙	1 tsp rice wine

煮鴨香料 | Spices for cooking duck

陳皮 2 片	2 slices dried tangerine peel
八角 2 粒	2 octagonals
薑 2 片	2 slices ginger
葱 1 棵	1 stalk spring onion
清水 2 杯	2 cups water

4~5 人
Serves 4~5

2~3 小時
2~3 hours

芡汁 | Gravy

煮鴨汁 1/2 杯	1/2 cup sauce of cooking duck
生粉 1 茶匙	1 tsp cornstarch
糖 1 茶匙	1 tsp sugar
麻油少許	Some sesame oil
胡椒粉少許	Pinch of pepper
清水 2 湯匙	2 tbsps water

做法 | Method

1. 鴨洗淨拭乾，把調味料掃勻鴨身。
2. 荔芋去皮，切角。燒熱油鍋，用中火炸至微黃。
3. 用沸水沖去鴨身上的油，放入鍋中，加入煮鴨香料煮滾，轉小火煮 1 1/2 小時，熄火，待涼後取出切件。
4. 燒熱 1 湯匙油，爆香薑片，倒入荔芋，注入 1 杯清水煮 15-20 分鐘，至荔芋開始腍軟。
5. 放入芡汁煮 5 分鐘，將鴨件放上即可。

1. Rinse duck and wipe dry, sweep seasonings onto the body of duck evenly.
2. Peel and cut taro into wedges. Heat a wok of oil, deep-fry taro over medium heat till yellowish.
3. Remove grease on the body of duck by rinsing with boiling water. Put the duck into the pot, add spices for cooking duck and bring to a boil. Turn to low heat and simmer for 1 1/2 hours, turn off heat, let cool and cut into pieces.
4. Heat 1 tbsp of oil, sauté ginger slices, add taro and a cup of water and cook for 15 to 20 minutes until taro becomes tender.
5. Add gravy and cook for 5 minutes, add duck pieces. Serve.

入廚貼士 | Cooking Tips

- 沒有芋頭，可以只煮鴨，效果也不錯。
- If taro is not added, the duck alone could also be delicious.

冬菇海參燜鴨掌

Stew Duck's Feet with Dried Black
Mushrooms and Sea Cucumber

材料 | Ingredients

急凍鴨掌 6-8 隻
急凍海參 300 克（已浸發）
五花腩 300 克
冬菇 5-6 朵（浸軟）
薑 2-3 片
蒜子 2-3 粒
上湯 1 杯

6-8 frozen duck's feet
300 g frozen sea cucumber (soaked)
300 g pork belly
5-6 dried black mushrooms (soaked)
2-3 slices ginger
2-3 cloves garlic
1 cup chicken stock

Stew・
Chicken
& Duck

醃料	Marinade		調味料	Seasonings
糖 1 茶匙	1 tsp sugar	糖 2 茶匙	2 tsps sugar
薑汁酒 1 茶匙	1 tsp ginger wine	蠔油 2 茶匙	2 tsps oyster sauce
生粉 1 茶匙	1 tsp cornstarch	生粉 1 茶匙	1 tsp cornstarch
油 1 茶匙	1 tsp oil	鹽 1/4 茶匙	1/4 tsp salt
鹽 1/2 茶匙	1/2 tsp salt	清水 1/3 杯	1/3 cup water

做法 | Method

1. 急凍鴨掌和海參分別汆水，過冷河。

2. 五花腩洗淨，切塊，加入醃料拌勻，待 10 分鐘。燒熱鍋，下五花腩炒至半熟，盛起。

3. 熱鍋下 1-2 湯匙油，放薑片和蒜子爆香，加入冬菇、海參和五花腩略炒，濆酒，注入上湯煮 30 分鐘。

4. 倒入鴨掌煮 10 分鐘，下調味料煮至濃稠收汁便可。

1. Blanch frozen duck's feet and sea cucumber respectively, rinse and drain.

2. Rinse pork belly, cut into pieces, add marinade and leave for 10 minutes. Heat a wok and stir-fry until half done and dish up.

3. Heat wok with 1-2 tbsps of oil, sauté ginger slices and garlic until fragrant, add dried black mushrooms, sea cucumber and pork belly and stir-fry for a while, drizzle wine, pour in chicken stock and cook for 30 minutes.

4. Add duck's feet and cook for 10 minutes, add seasonings and cook until sauce thickens. Serve.

入廚貼士 | Cooking Tips

- 急凍鴨掌和海參汆水，可加點薑汁酒同煮，去除雪藏味。
- When blanching frozen duck's feet and sea cucumber, add some ginger wine could help to remove frozen flavor.

4~6 人
Serves 4~6

50 分鐘
50 minutes

麥芽糖燜一字排

Stewed Spare Ribs with Maltose

材料 | Ingredients

一字排 600 克	600 g spare ribs
乾葱 4 粒	4 shallots
薑 3 片	3 slices ginger
八角 2 粒	2 star anise
麥芽糖 3 湯匙	3 tbsps maltose
紹酒 2 湯匙	2 tbsps Shaoxing wine
水 3 杯	3 cups water

醃料 | Marinade

喼汁 4 湯匙	4 tbsps Worcestershire sauce
老抽 2 湯匙	2 tbsps dark soy sauce

做法 | Method

1. 一字排洗淨，抹乾水分，加入醃料醃片刻。
2. 燒熱鍋，下少許油，加入排骨煎至兩面微黃。加入薑片及乾葱，潽酒，加入餘下醃料及八角拌勻。
3. 注入滾水 3 杯煮滾，轉文火爛 40 分鐘至排骨軟腍及汁液濃稠。
4. 加入麥芽糖拌勻，待排骨入味即成。

1. Rinse spare ribs, drain and marinate for a while.
2. Heat a wok with some oil, pan-fry spare ribs until both sides are slightly yellow. Add ginger slices and shallots, sizzle wine, mix well with remaining marinade and star anise.
3. Add 3 cups of boiling water and bring to a boil. Turn to low heat and stew for 40 minutes until spare ribs are soft and sauce is thickened.
4. Add maltose and mix well, cook until spare ribs are well flavored. Serve.

馬鈴薯燜排骨

Stewed Spare Ribs with Potatoes

材料 | Ingredients

排骨 300 克
馬鈴薯 2-3 個（切角）
薑茸 1 茶匙
蒜茸 1 茶匙
清水 1 杯

300 g spare ribs
2-3 potatoes (cut into wedges)
1 tsp minced ginger
1 tsp minced garlic
1 cup water

4~6 人
Serves 4~6

15~20 分鐘
15~20 minutes

醃料 | Marinade

薑汁酒 2 茶匙	2 tsps ginger wine
油 2 茶匙	2 tsps oil
生粉 1 茶匙	1 tsp cornstarch
糖 1 茶匙	1 tsp sugar
鹽 1/2 茶匙	1/2 tsp salt

調味料 | Seasonings

糖 2 茶匙	2 tsps sugar
鹽 1/4 茶匙	1/4 tsp salt

做法 | Method

1. 排骨洗淨，抹乾水分，加入醃料醃 20 分鐘，備用。
2. 燒熱鍋，將馬鈴薯用油炸至金黃，撈起，瀝油。
3. 燒熱鍋，下 1/2 湯匙油，放入薑茸和蒜茸爆香，放入排骨炒香；注入清水煮 10-15 分鐘。
4. 倒入已炸的薯角繼續煮 5 分鐘，下調味料拌勻，上碟。

1. Rinse spare ribs and pat dry, marinate for 20 minutes for later use.
2. Heat wok and deep-fry potatoes until golden, dish up and drain.
3. Heat wok with 1/2 tbsp of oil, sauté mined ginger and garlic until fragrant, add spare ribs and stir-fry until fragrant. Add water and cook for 10-15 minutes.
4. Add cooked potato wedges and cook for another 5 minutes, add seasonings and stir well. Serve.

入廚貼士 | Cooking Tips

- 排骨帶點肥肉的話，燜後會脆軟帶肉汁。
- The dish will be full of gravy if spare ribs with a little fat are used.

涼瓜燜排骨

Stewed Spare Ribs with Bitter Gourd

材料 | Ingredients

排骨 300 克
涼瓜 1 個
薑茸 1 茶匙
蒜茸 1 茶匙
清水 1 杯

300 g spare ribs
1 bitter gourd
1 tsp minced ginger
1 tsp minced garlic
1 cup water

⦿⦿⦿ 醃料 | Marinade

薑汁酒 2 茶匙	2 tsps ginger wine
油 2 茶匙	2 tsps oil
生粉 1 茶匙	1 tsp cornstarch
糖 1 茶匙	1 tsp sugar
鹽 1/2 茶匙	1/2 tsp salt

入廚貼士 | Cooking Tips

- 涼瓜用鹽拌勻後再汆水，可降低苦味。
- Mix bitter gourd with salt and then blanch can reduce the bitterness.

⦿⦿⦿ 調味料 | Seasonings

糖 2 茶匙
生粉 1 茶匙
鹽 1/2 茶匙
老抽 1/2 茶匙
鹽 1/4 茶匙
清水 3 湯匙

2 tsps sugar
1 tsp cornstarch
1/2 tsp salt
1/2 tsp dark soy sauce
1/4 tsp salt
3 tbsps water

⦿⦿⦿ 做法 | Method

1. 涼瓜去籽，切塊，用鹽拌勻後待片刻，洗淨。燒熱鍋，用少量油炒涼瓜片刻，盛起。
2. 排骨洗淨，瀝乾，切塊。加入醃料拌勻。燒熱鍋，炒至半熟，盛起。
3. 燒熱鍋，下 1/2 湯匙油，放薑茸和蒜茸爆香，倒入排骨和苦瓜炒數下，灒酒，倒入清水煮 15-20 分鐘至腍軟。
4. 下調味料煮至濃稠收汁，上碟。

1. Seed bitter gourd and cut into pieces, mix with some salt and leave for a while, rinse. Heat wok and stir-fry bitter gourd with some oil for a while, dish up.
2. Rinse and drain spare ribs and cut into pieces. Marinate and mix well. Heat wok and stir-fry until half done. Dish up.
3. Heat wok with 1/2 tbsp of oil, sauté minced ginger and garlic until fragrant, add spare ribs and bitter gourd and stir-fry for a while. Drizzle wine, pour in water and cook for 15-20 minutes until soft.
4. Add seasonings and cook until sauce thickens. Serve.

蜜汁金沙骨

Braised Spare Ribs in Honey Sauce

⦿⦿⦿ 材料 | Ingredients

長形排骨 4-6 條（每條約 4 吋長）
蜜糖 3 湯匙
老抽 1 湯匙

4-6 long-shaped spare ribs
(4 inches long each)
3 tbsps honey
1 tbsp dark soy sauce

4~6 人
Serves 4~6

20~30 分鐘
20~30 minutes

醃料 | Marinade

醬油 2 湯匙	2 tbsps soy sauce
糖 2 湯匙	2 tbsps sugar
生粉 1 湯匙	1 tbsp cornstarch
紅酒 1 湯匙	1 tbsp red wine

蜜汁 | Honey Sauce

蜜糖 2 湯匙	2 tbsps honey
老抽 1 茶匙	1 tsp soy sauce
清水 2 湯匙	2 tbsps water

做法 | Method

1. 將金沙骨用醃料醃 3 小時。
2. 用錫紙包好，再蒸 1/2 小時，將醬汁和金沙骨分別盛起，備用。
3. 燒熱易潔鑊，放少許油，再把金沙骨回鑊，以慢火煎片刻。
4. 倒進蜜汁續以慢火煎至醬汁收乾，上碟。

1. Marinate spare ribs for 3 hours.
2. Wrap spare ribs with an aluminum foil, steam for 1/2 hour, dish up the sauce and the spare ribs respectively.
3. Heat some oil in a non-stick wok, return spare ribs and pan-fry for a while.
4. Pour in honey and continue to cook over low heat and simmer till sauce dries up. Serve.

入廚貼士 | Cooking Tips
- 採用先蒸後煎封，再與醬汁爛煮的方法，確保排骨肉腍而乾爽。
- The method of steam, then pan-fry and braise with the sauce could ensure that the spare ribs are soft but not having too much sauce.

豉汁金瓜肉排

Stewed Spare Ribs and Pumpkin in Black Bean Sauce

材料 | Ingredients

南瓜 600 克
排骨 300 克
薑茸 1 茶匙
蒜茸 1 茶匙
豆豉 1 茶匙
清水 1 杯

600 g pumpkin
300 g spare ribs
1 tsp minced ginger
1 tsp minced garlic
1 tsp fermented black beans
1 cup water

醃料 | Marinade

薑汁酒 2 茶匙	2 tsps ginger wine
油 2 茶匙	2 tsps oil
生粉 1 茶匙	1 tsp cornstarch
糖 1 茶匙	1 tsp sugar
鹽 1/2 茶匙	1/2 tsp salt

調味料 | Seasonings

糖 2 茶匙
生粉 1 茶匙
老抽 1/2 茶匙
鹽 1/4 茶匙
清水 3 湯匙

2 tsps sugar
1 tsp cornstarch
1/2 tsp dark soy sauce
1/4 tsp salt
3 tbsps water

入廚貼士 | Cooking Tips

- 南瓜的皮很硬，可先用菜刀逐少切掉，比較容易處理。
- Since the skin of pumpkin is very hard, it would be relatively easy to handle if it is cut with a kitchen knife bit by bit.

做法 | Method

1. 南瓜去籽和削皮，切塊，洗淨。燒熱鍋，用少量油炒片刻。
2. 排骨洗淨，切塊，加入醃料撈勻。燒熱鍋，下排骨炒至半熟。
3. 熱鑊下 1-2 湯匙油，放薑茸、蒜茸和豆豉爆香，倒入排骨和南瓜炒數下，潶酒，倒入清水煮 15-20 分鐘至腍軟。
4. 下調味料煮至濃稠收汁，上碟。

1. Seed and peel pumpkin, cut into pieces, rinse. Heat a wok and stir-fry with some oil for a while.
2. Rinse spare ribs and cut into pieces, mix with marinade. Heat a wok and stir-fry until half done.
3. Heat a wok with 1-2 tbsps of oil, sauté minced ginger, minced garlic and fermented black beans until fragrant, add spare ribs and pumpkin and stir-fry for a while. Drizzle wine, add water and cook for 15-20 minutes until soft.
4. Add seasonings and cook until sauce thickens. Serve.

慈菇燜豬肉

Stewed Pork with Arrowhead

◯◯ 材料 | Ingredients

五花腩 300 克
慈菇 6-8 個
薑 2-3 片
中國蒜 1 條（切段）
蒜子 1-2 粒
南乳 1/2 磚
片糖 1/2 片
清水 1 杯

300 g pork belly
6-8 arrowheads
2-3 slices ginger
1 Chinese garlic (sectioned)
1-2 cloves garlic
1/2 piece red beacurd
1/2 piece slab sugar
1 cup water

4~6 人
Serves 4~6

30~40 分鐘
30~40 minutes

醃料	Marinade	

糖 1 茶匙　　1 tsp sugar
薑汁酒 1 茶匙　1 tsp ginger wine
生粉 1 茶匙　　1 tsp cornstarch
油 1 茶匙　　1 tsp oil
鹽 1/2 茶匙　　1/2 tsp salt

調味料 | Seasonings

生粉 1 茶匙　　1 tsp cornstarch
鹽 1/4 茶匙　　1/4 tsp salt
清水 1/3 杯　　1/3 cup water

做法 | Method

1. 慈菇洗淨，用小刀刮去皮，切角。燒熱鍋，用少量油炒片刻。
2. 五花腩洗淨，切塊，加入醃料撈勻。燒熱鍋，炒至半熟。
3. 燒熱鍋，下 1-2 湯匙油，放薑片、蒜子和南乳爆香，倒入五花腩和慈菇炒數下，灒酒，倒入清水和片糖煮 20 分鐘至腍軟。
4. 下調味煮至汁液濃稠，加入中國蒜炒勻，上碟。

1. Rinse arrowheads and peel with a small knife, cut into wedges. Heat wok and stir-fry with some oil for a while.
2. Rinse pork belly and cut into pieces, add marinade and mix well. Heat wok and stir-fry until half done.
3. Heat wok with 1-2 tbsps of oil, sauté ginger slices, garlic and red beancurd until fragrant, add pork belly and arrowheads and stir-fry for a while, drizzle wine, pour in water and slab sugar and cook for 20 minutes until soft.
4. Add seasonings and cook until sauce thickens, add Chinese garlic and stir well. Serve.

入廚貼士 | Cooking Tips

- 慈菇是時令食物，所以不是常常買到，改用芋頭或准山都可以。
- Arrowhead is a seasonal ingredient and may not be easily bought, it can be replaced by taro or yam.

冬菇蠔豉燜豬手

Stewed Trotter with Dried Black Mushrooms and Dried Oysters

◯◯◯ **材料｜Ingredients**

蠔豉 6-8 隻　　　　　蒜子 2-3 粒
豬手 1 隻　　　　　　濕髮菜少許
冬菇 10-12 朵（浸軟）上湯 1 杯
薑 2-3 片

6-8 dried oysters
1 trotter
10-12 dried black mushrooms (soaked)
2-3 slices ginger
2-3 cloves garlic
black moss (soaked)
1 cup chicken stock

 調味料 | Seasonings

糖 2 茶匙	2 tsps sugar
蠔油 2 茶匙	2 tsps oyster sauce
生粉 1 茶匙	1 tsp cornstarch
鹽 1/4 茶匙	1/4 tsp salt
清水 1/3 杯	1/3 cup water

 做法 | Method

1. 蠔豉洗淨，燒熱一鍋水，蠔豉加 2 茶匙薑汁酒，以大火隔水蒸 10 分鐘。

2. 豬手刮去毛，洗淨。燒熱一鍋水，豬手放沸水中焯煮 20 分鐘，不用蓋蓋。取出過冷，洗淨。

3. 燒熱鍋，下 1-2 湯匙油，放薑片和蒜子爆香，加入冬菇、蠔豉和豬手炒數下，潷酒，注入上湯煮 30 分鐘。

4. 下調味料和髮菜煮至汁液濃稠便可。

1. Rinse dried oysters. Heat a wok of water and steam oysters with 2 tsps of ginger wine for 10 minutes.

2. Shave trotter and rinse. Heat a wok of water and cook trotter for 20 minutes without cover the lid, remove and rinse with cold water.

3. Heat a wok with 1-2 tbsps of oil, sauté ginger slices and garlic until fragrant, add dried black mushrooms, dried oysters and trotter and stir-fry for a while. Drizzle wine, pour in chicken stock and cook for 30 minutes.

4. Add seasonings and black moss and cook until sauce thickens. Serve.

入廚貼士 | Cooking Tips
- 冬菇用生粉撈勻，沖洗乾淨，擠乾水分，可去掉泥味。
- To get rid of the mud smell of dried black mushrooms, mix with cornstarch, rinse and squeeze excess water.

南乳蓮藕燜豬手

4~6 人
Serves 4~6

40~50 分鐘
40~50 minutes

◯◯ 材料 | Ingredients

豬手 1 隻	1 trotter
蓮藕 600 克	600 g lotus roots
薑 2-3 片	2-3 slices ginger
蒜頭 2-3 粒	2-3 cloves garlic
南乳 3 湯匙	3 tbsps red beancurd
清水 4 杯	4 cups water

◯◯ 調味料 | Seasonings

片糖 1 塊
鹽 1/4 茶匙

1 slice slab sugar
1/4 tsp salt

◯◯ 做法 | Method

1. 豬手解凍。燒熱一鍋水，豬手放沸水中焯煮 20 分鐘，不用蓋蓋。取出過冷，洗淨。

2. 蓮藕用小刀刮皮，切塊。

3. 燒熱鍋，下 1-2 湯匙油，放入薑片、南乳和蒜頭爆香，放入豬手和蓮藕炒透，注入清水煮至脍軟。

4. 加入調味料，煮至汁液濃稠，上碟。

1. Thaw trotter. Heat a wok of water and cook trotter for 20 minutes without cover the lid, remove and rinse with cold water.

2. Peel lotus roots with a small knife, cut into pieces.

3. Heat a wok with 1-2 tbsps of oil, add ginger slices, red beancurd and garlic, add trotter and lotus root and stir-fry thoroughly, add water and cook until soft.

4. Add seasonings and cook until sauce dries up. Serve.

入廚貼士 | Cooking Tips

- 蓮藕用刀拍扁比刀切好，可避免氧化和保持原汁原味。
- It is better to pat lotus roots instead of cutting as it can prevent oxidation and maintains original flavor.

芋頭燜臘肉

Stew Preserved Pork Belly with Taro

⊙⊙⊙ 材料 | Ingredients

芋頭 600 克
臘肉 1/2 條（約 150 克）
蒜頭 2-3 粒
清水 2 杯

600 g taro
1/2 strips preserved pork belly (about 150 g)
2-3 cloves garlic
2 cups water

調味料 | Seasonings

椰汁 1/2 杯	1/2 cup coconut milk
鹽 1/2 茶匙	1/2 tsp salt
糖 1/2 茶匙	1/2 tsp sugar

做法 | Method

1. 臘肉洗淨，燒熱一鍋水，加入臘肉焯 1-2 分鐘，撈起，過冷，起皮，切件。
2. 芋頭去皮，切塊，洗淨。
3. 燒熱鍋，下 1-2 湯匙油，放入蒜頭和臘肉炒勻，再加入芋塊，注入清水煮至腍軟。
4. 加入調味料，煮至汁液濃稠，上碟。

1. Rinse preserved pork belly. Heat a wok of water and blanch preserved pork belly for 1-2 minutes, pick up and rinse with water, remove skin and cut into pieces.
2. Peel taro, cut into pieces, rinse.
3. Heat wok with 1-2 tbsps of oil, add garlic and preserved pork belly and stir well, then add taro pieces, add water and cook until soft.
4. Add seasonings, cook until sauce dries up. Serve.

入廚貼士 | Cooking Tips
- 臘肉用沸水焯煮，可去掉表面肥油，切塊時較省力。
- Blanch preserved pork belly in boiling water can remove the fat on surface, it also saves energy when cutting.

番茄燴肉丸

Stewed Meatballs with Tomatoes

材料 | Ingredients

番茄 600 克（切角）
免治豬肉 300 克
蒜茸 1 茶匙
薑茸 1 茶匙

600 g tomatoes (cut into wedges)
300 g minced pork
1 tsp minced garlic
1 tsp minced ginger

4~6 人
Serves 4~6

30 分鐘
30 minutes

醃料 | Marinade

糖 1 茶匙
生粉 1 茶匙
油 1 茶匙
鹽 1/2 茶匙
紹興酒 1/2 茶匙

1 tsp sugar
1 tsp cornstarch
1 tsp oil
1/2 tsp salt
1/2 tsp Shaoxing wine

調味料 | Seasonings

茄汁 1/3 杯
糖 1-2 湯匙
生粉 2 茶匙
鹽 1/4 茶匙
清水 1/4 杯

1/3 cup ketchup
1-2 tbsps sugar
2 tsps cornstarch
1/4 tsp salt
1/4 cup water

做法 | Method

1. 免治豬肉放大碗中，加醃料拌勻，再大力攪拌，置冰箱待 20 分鐘。
2. 燒熱油一鍋至八成滾，把肉丸放入油中炸至金黃，約六至七成熟，瀝油。
3. 燒熱鍋，下 1-2 茶匙油，放入蒜茸和薑茸爆香，倒入番茄角和豬肉丸煮至濃稠，下調味料煮至汁液濃稠便可。

1. Marinate minced pork and mix well in a bowl, then stir vigorously. Put into the freezer for 20 minutes.
2. Heat a wok of oil to 80% boil, deep-fry meatballs until 60-70% done.
3. Heat wok with 1-2 tsps of oil, sauté minced garlic and ginger until fragrant, add tomatoes and meatballs and cook until sauce thickens. Add seasonings and cook until sauce dries up. Serve.

入廚貼士 | Cooking Tips
- 肉丸調味後需要放冰箱內冷藏，肉質會較有彈性。
- The meatballs will be elastic if it is stored in the freezer after seasoning.

韭菜燒腩燜豆腐

Stewed Beancurd with Leeks and
Roasted Pork Belly

材料 | Ingredients

燒腩 200 克
韭菜 150 克（切段）
板豆腐 1 磚
蒜頭 1 粒（略拍扁）
薑數片
甘筍數片
清水 1 杯

200 g roasted pork belly
150 g leeks (sectioned)
1 block firm beancurd
1 clove garlic (slightly pat)
a few slices of ginger
a few slices of carrots
1 cup water

⫯⫯ 芡汁 | Thickening

糖 1 茶匙	1 tsp sugar
生粉 1 茶匙	1 tsp cornstarch
鹽 1/2 茶匙	1/2 tsp salt
老抽 1/2 茶匙	1/2 tsp dark soy sauce
清水 3 湯匙	3 tbsps water

⫯⫯ 做法 | Method

1. 板豆腐切件，燒熱鍋，下 1-2 湯匙油煎至金黃，盛起瀝油。

2. 燒熱鍋，下 1-2 茶匙油，爆香蒜頭和薑片，放入燒腩和豆腐炒香，注入清水煮至腍軟。

3. 下芡汁煮至濃稠，再加入韭菜和甘筍煮片刻，上碟。

1. Cut firm beancurd into pieces. Heat wok with 1-2 tbsps of oil, pan-fry beancurd until golden brown, dish up and drain.

2. Heat a wok with 1-2 tsps of oil, sauté garlic and ginger slices until fragrant, add roasted pork belly and beancurd and stir-fry, add water and cook until soft.

3. Add thickening sauce and cook until thickens, add leeks and carrot and cook for a while, serve.

入廚貼士 | Cooking Tips
- 韭菜不要煮得太熟，否則會變黃而影響味道。
- Do not cook leeks for too long, otherwise it will turn yellow and the taste is not good.

蘿蔔燜牛腩

Stewed Beef Brisket with Turrip

4~6 人
Serves 4~6

45 分鐘
45 minutes

材料 | Ingredients

牛腩 500 克
白蘿蔔 400 克
葱 1 條
八角 1 粒
水 8 杯

500 g beef brisket
400 g turnip
1 stalk spring onion
1 star anise
8 cups water

調味料 | Seasonings

生抽 2 茶匙
酒 1 茶匙
糖 1/2 茶匙
鹽 1/4 茶匙

2 tsps soy sauce
1 tsp wine
1/2 tsp sugar
1/4 tsp salt

芡汁料 | Thickening

生粉 1 茶匙
水 1 湯匙

1 tsp cornflour
1 tbsp water

烟 · 豬牛類

Stew ·
Pork & Beef

做法 | Method

1. 牛腩洗淨，切小塊，汆水，沖淨泡沫，瀝乾。

2. 白蘿蔔去皮，洗淨後切小塊；葱洗淨，切段。

3. 燒熱一鍋水，下牛腩和八角煮滾，改慢火煮 30 分鐘。加入白蘿蔔同煮，下調味料煮至湯汁濃稠，勾芡，撒上葱段，即可。

1. Rinse beef brisket and cut into small pieces. Blanch, rinse and drain.

2. Peel turnip, rinse and cut into small pieces. Rinse spring onion and cut into sections.

3. Boil a pot of water, add beef brisket and star anise. Turn to low heat and cook for 30 minutes. Add turnip and seasonings and cook until the broth nearly dries up. Thicken and sprinkle with spring onion.

柱侯牛筋

Stewed Beef Tendon in Chu Hou paste

⚬⚬⚬ 材料 | Ingredients

急凍牛筋 300 克
白蘿蔔 300 克
薑 3-4 片
蒜頭 1-2 粒
中國大蒜 1 條
清水 2 杯

300 g frozen beef tendon
300 g turnip
3-4 slices ginger
1-2 cloves garlic
1 Chinese garlic
2 cups water

調味料 │ Seasonings

柱侯醬 1 湯匙	1 tbsp Chu Hou paste
糖 1 茶匙	1 tsp sugar
鹽 1 茶匙	1 tsp salt
生粉 1 茶匙	1 tsp cornstarch
清水 3 湯匙	3 tbsps water

做法 │ Method

1. 牛筋解凍。燒熱一鍋水，放入薑片和蒜頭，加入牛筋焯煮 5 分鐘，取出過冷，瀝乾。
2. 蘿蔔去皮，切角。中國大蒜洗淨，切段。
3. 燒熱鍋，下 1-2 茶匙油，爆香薑片、蒜頭和柱侯醬爆香，倒入白蘿蔔和牛筋炒透，潷酒。注入清水煮 30 分鐘。
4. 下調味料煮至湯汁濃稠，盛起。

1. Thaw beef tendon. Heat a wok of water, add ginger slices, garlic and beef tendon and blanch for 5 minutes, remove and drain, leave to cool.
2. Peel turnip and cut into wedges. Rinse Chinese garlic, cut into sections.
3. Heat a wok with 1-2 tsps of oil, sauté ginger slices, garlic and Chu Hou paste until fragrant, add turnip and beef tendon and stir-fry until fragrant, drizzle wine. Add water and cook for 30 minutes.
4. Add seasonings and cook until sauce dries up, serve.

入廚貼士 │ Cooking Tips

- 牛筋煮完後熄火，不要揭蓋，待 10 分鐘，效果更好。
- The result of beef tendon will be better if switch off heat after cooking, do not open the lid and wait for 10 minutes.

咖喱牛筋腩

Curry Beef Tendon and Beef Brisket

材料 | Ingredients

熟牛筋腩 450 克
洋葱 1 個（切角）
三色椒各 1/4 個（切角）
甘筍 150 克（切角）
馬鈴薯 2-3 個（切角）

薑茸 1 茶匙
蒜茸 1 茶匙
辣椒茸 1 茶匙
清水 1 杯

450 g cooked beef tendon and beef brisket
1 onion (cut in wedges)
1/4 assorted color of bell peppers (cut in wedges)
150 g carrot (cut in wedges)
2-3 potatoes (cut in wedges)
1 tsp minced ginger
1 tsp minced garlic
1 tsp minced red chili
1 cup water

4~6 人
Serves 4~6

20~25 分鐘
20~25 minutes

醃料 | Marinade		調味料 | Seasonings	
薑汁酒 2 茶匙	2 tsps ginger wine	油咖喱醬 2 茶匙	2 tsps curry paste
油 2 茶匙	2 tsps oil	糖 2 茶匙	2 tsps sugar
生粉 1 茶匙	1 tsp cornstarch	咖喱粉 1 茶匙	1 tsp curry powder
糖 1 茶匙	1 tsp sugar	鹽 1/4 茶匙	1/4 tsp salt
鹽 1/2 茶匙	1/2 tsp salt		

做法 | Method

1. 牛筋腩解凍，用少許鹽擦洗，沖淨，抹乾水分，加入醃料醃 1 小時備用。
2. 燒熱鍋，將洋葱、馬鈴薯、三色椒、甘筍用油炸至金黃，撈起，瀝油。
3. 熱鑊下 1-2 湯匙油，放入薑茸、蒜茸、辣椒茸、油咖喱醬和咖喱粉爆香，放入牛筋腩和清水煮 15 分鐘。
4. 倒入已炸的雜菜繼續煮 5 分鐘，下調味料拌勻，上碟。

1. Thaw beef tendon and beef brisket, scrub with some salt, rinse and pat dry. Marinate for 1 hour.
2. Heat a wok of oil, deep-fry onion, potatoes, assorted color of bell peppers and carrot until golden, dish up and drain.
3. Heat a wok with 1-2 tbsps of oil, sauté minced ginger, garlic, red chili, curry paste and curry powder until fragrant. Add beef tendon, beef brisket and water and cook for 15 minutes.
4. Add cooked mixed vegetables and continue to cook for 5 minutes, add seasonings and mix well, serve.

入廚貼士 | Cooking Tips
- 牛筋腩可預先用高速煲或已煲湯的牛腩作材料。
- Beef tendon and beef brisket could be cooked in advance by a pressure cooker or cooked for soup.

爛出美味

編著
何美好

編輯
紫彤

美術設計
Venus

排版
劉葉青

出版者
萬里機構出版有限公司
香港鰂魚涌英皇道1065號東達中心1305室
電話：2564 7511
傳真：2565 5539
電郵：info@wanlibk.com
網址：http://www.wanlibk.com
　　　http://www.facebook.com/wanlibk

發行者
香港聯合書刊物流有限公司
香港新界大埔汀麗路36號
中華商務印刷大廈3字樓
電話：2150 2100
傳真：2407 3062
電郵：info@suplogistics.com.hk

承印者
美雅印刷製本有限公司

出版日期
二零一八年十月第一次印刷

萬里機構

萬里 Facebook